RAM PUBLICATIONS

Treasure Recovery from Sand and Sea
Charles Garrett shares a lifetime of experience to explain how any treasure hunter can reach the "blanket of wealth" beneath sands nearby and under the world's oceans, lakes and streams.

Buried Treasure of the United States
Complete field guide for finding treasure includes state-by-state listing of sites where treasure may be found.

Weekend Prospecting
Explains how to enjoy holidays and vacations by prospecting with metal detectors and gold pans.

Successful Coin Hunting
World's most authoritative guide on where and how to find valuable coins with a metal detectors.

Modern Metal Detectors
THE recognized text for hobby and professional metal detecting provides expertise in all fields of detector use.

Detector Owner's Field Manual
Explains total capabilities of and tells how to use any type of metal detector.

Treasure Hunting Pays Off
An introduction to all facets of the hobby of treasure hunting with a metal detector.

VLF Metal Detector Handbook
Tells how to understand, select and use the modern VLF high performance VLF metal detector.

Electronic Prospecting
Explains how anyone can use metal detectors to find gold and silver veins and nuggets.

Gold Panning is Easy
Excellent field guide illustrates how to find and pan gold as quickly and easily as a professional.

Treasure Hunter's Manual #6
Quickly guides the beginner through the mysteries of full-time treasure hunting.

Treasure Hunter's Manual #7
Begins where #6 ends to explain professional methods of research, recovery and disposition of treasure.

Garrett Guides to Treasure

You Can Find An Ounce of Gold a Day

Metal Detectors Can Help You Find Wealth

You Can Find Wealth on the Beach

Find Wealth in the Surf

WEEKEND PROSPECTING

RAM BOOKS

ROY LAGAL

RAM BOOKS

ISBN 0-915920-48-4
Library of Congress Catalog Card No. 88-060012
Weekend Prospecting

First Edition Printing, January 1988

88 89 90 10 9 8 7 6 5 4 3 2 1

By ROY LAGAL

Weekend Prospecting
Find an Ounce of Gold a Day
Gold Panning is Easy
Detector Owner's Field Manual

(With Charles Garrett)
The Complete VLF-TR Metal Detector Handbook

(With Charles Garrett and Bob Grant)
Electronic Prospecting

CONTENTS

Preface

There is no deep mystery to prospecting for gold. But, there is excitement. And, there can be great profit. No large expenditures of cash are required. Success is limited only by the enterprise and dreams of the prospector.

WEEKEND PROSPECTING is a basic instruction guide for the beginner, and it is designed to help the professional in areas where expertise or knowledge is lacking. Assisted by international experts, the Author compiled this guide while making the instructional video production, "Weekend Prospecting." It contains a considerable body of knowledge that would be difficult for any one person to accumulate in a lifetime. While the book should be of value to even the most experienced prospector, it is designed to serve as a simple guide for the beginner.

Scores of publications, technical manuals and instruction books have been written by experienced prospectors on various phases of prospecting and mining. Most are detailed and quite thorough on a single subject. The picture they present is often too complex for a beginner. I know of no publication that compiles information on various phases of the field in a simple, beginner's guidebook.

After developing an interest in prospecting, an individual will wish to study some of these more detailed books in specific areas of interest.

It is my hope with WEEKEND PROSPECTING, however, that efforts of experts to combine their knowledge in one condensed guide will benefit both beginner and professional. Remember, knowledge is always the key to success.

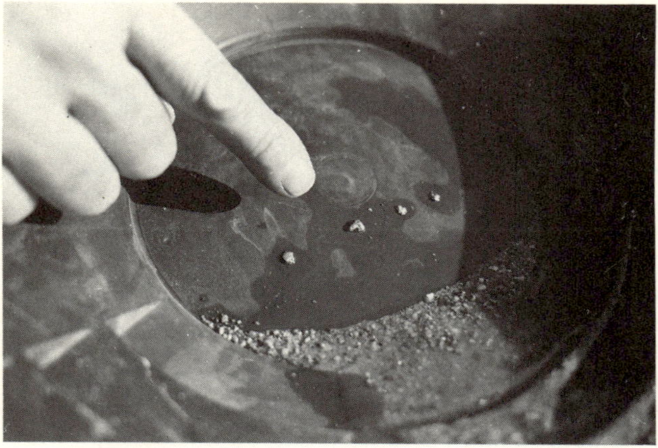

Four jewelry-size gold nuggets (above) were panned from the Snake River of Idaho by Roy Lagal. Silver recovered by the author from a small "glory hole" along the fabled Batopillas River in Mexico is shown in the Gravity Trap gold pan below.

Introduction

WHY SEARCH FOR GOLD?

The individual selects a hobby for many different reasons—enjoyment, outdoor activity, adventure, physical exercise, monetary profit. Weekend prospecting combines all of these, with emphasis on potential profits. Joy, happiness and other spiritual benefits are immeasurable, but monetary rewards come in sizes ranging from small nuggets worth a few hundred dollars to the huge "Hand of Faith" nugget that was sold by a beginning prospector for one million dollars.

Amateurs and professionals alike can discover gold nuggets with a metal detector, find gold in a pan or dredge for gold. They can relive the past, participate in adventure in the great outdoors and enjoy physical exercise...all the while with the potential of monetary profit at the risk of only a few dollars.

WHERE DO YOU FIND GOLD?

Gold has been found in almost every state of the United States and every country in the world. And, just because gold has not been found in an area does not mean the area is barren. Perhaps nature's treasure is merely waiting to be discovered. It is best, however, for beginners to prospect known mineral-producing areas because it is here they will most likely find gold and obtain needed expertise. Asking a few questions and carrying out a little research should reveal such areas. On-site experience will produce knowledge and confidence that leads to success.

HOW DO YOU FIND GOLD?

This book will get you started by carefully explaining basic tools of prospecting. You will learn to pinpoint gold locations as you develop "sniping" techniques. You can start with a gold pan, then multiply your successes with a metal detector, dredge, dry washer or other more sophisticated tools and equipment.

You can become a professional prospector, or you can fill your weekends and holidays with the joy of discovering the earth's wealth alongside family and friends. You can begin by working at your own speed, but heed this warning: Once you experience the thrill of finding gold...once you taste success, a fire will be ignited within you. Don't try to put it out because you can't. You'll be eager to discover your share of nature's gold as millions of others have. You will learn about striking it rich in the gold fields!

Chapter 1

TRUE STORIES OF SUCCESS

This chapter contains tales of gold discoveries made by beginners and professionals who applied modern prospecting methods. Published photographs seem to contain a great deal of "found" gold, but they can show only a fraction of the finds that have been made. Television occasionally presents news of a discovery along with information on an area. Newspapers and magazines carry articles relating to gold strikes. Treasure and prospecting publications are filled with details of finds. The more you become involved with prospecting, the greater will be your amaze-

Weekend prospectors can discover gold nuggets with modern metal detectors. The computerized Garrett Grand Master Hunter, shown here, is microprocessor controlled and is especially qualified for electronic prospecting and ore sampling.

ment at successes achieved in the gold fields and the potential for rich discoveries. You'll learn first-hand of some and hear many more success stories that you know to be true.

Rising gold prices have expanded the use of modern methods into new locales in foreign countries. Millions of dollars in gold is being discovered by amateurs every year, with these discoveries factually documented.

Does this mean we should all quit our jobs and rush to the gold fields to get rich? Not by a long shot! Millions of dollars in gold is being discovered, but a great many people are looking...and, most are not particularly lucky. Weekend prospecting should be regarded as a fascinating, interesting and exciting hobby. It is all of that. But profit must be regarded only as a fringe benefit.

Consider the fisherman who spends all day in the cold rain, yet catches no fish. He may be momentarily discouraged but will return next week or next month trying to hook the big one. Similarly, the hobbyist who is patient and appreciates outdoor excercise will enjoy weekend prospecting. And, who knows? Next week...next year...maybe, tomorrow...the weekend prospector will be in front of television cameras or in a treasure magazine describing discovery of a huge nugget.

It could happen to you!

Chapter 2

BASIC TOOLS

Basic, inexpensive tools eventually needed by presistent and successful weekend prospectors include:

- Gold panning kit
- Gold suction bottle
- Waders (or rubber boots)
- Magnifying glass
- Tweezers
- Small can
- Ore bag or sack
- Underwater face mask & snorkel
- Screwdriver (or small rod for crevicing)
- Rock hammer
- Shovel
- Large and small pry bars
- VLF metal detector
- Lightweight 2-inch gold dredge

Equip yourself with the tools you need to get started. Then, learn in the field how to use them. As you gain experience and enjoy success, you will add additional and more sophisticated equipment. All of the tools in the above list are used by prospectors who search for gold in both wet and dry areas. All of these tools are described in this chapter and were used in this book's companion outdoor recreational video, "Weekend Prospecting."

GOLD PAN

Every tool is optional except the gold pan. It or a similar basin bowl is the most highly desired prospecting tool available. Whether you examine placer gold or a hard rock specimen ground into dust for identification, some type of pan must be used to separate the lightweight material from heaver concentrates.

Remember the old prospector from the Western movies riding his decrepit burro? All he had was a skillet or pie pan with which he scooped up gravel and "panned" for gold. As far as the "pan" is concerned, this depiction was more or less accurate although the prospector probably had some type of digging tool as well.

Pan designs have improved greatly since the old prospector's time. Today's gold pan is lighter in weight and offers greater speed in testing and classifying concentrates. It is also easier to handle and provides safer, surer results, especially for the beginner. Absolutely no experience is required for a beginner to enjoy success in the first panning session.

Careful attention should be given to the choice of a gold pan. They vary in shape, color and material. Today's most popular gold pan is made of plastic, which has significant advantages over tin or steel. Plastic is lightweight and is not affected by acid or other substances that can attack metal. Plastic is not attracted by magnets and is easy to cast or mold into a specific design. Non-biodegradable plastic will last for centuries while steel pans rust away.

Gold pans may be round, square or oblong. The round shape is preferred since it offers more variations in shaking, twirling or dumping. Plus, it also carries out the old-prospector stereotype, even if it is made of plastic.

This traditional shape has been improved with the addition of built-in riffles or traps. Many riffles seen in today's pans are of little help except to slow down the flow of water while panning. True gravity riffles are indented at a 90-degree right angle, offering a distinct advantage for the beginner. As professionals have learned how such gravity-type pans speed up the panning procedure, popularity of these pans has grown world-wide.

Color is also important. A light color similar to that of yellow gold will not show the gold readily enough for visual indentification. On the other hand, deep, dull colors make identification difficult by defusing or absorbing light, especially on cloudy days. Therefore, a color that contrasts with gold yet is not too dark is desired. From long experience I recommend a medium green color pan for the best identification of fine gold particles.

Gold pans are manufactured from five to 24 inches in size with the standard size approximately 14 inches. Finishing pans nine or ten inches in size make an excellent addition to the kit. They can be used to separate extremely fine gold from black sands and are light and small enough to be used by children, and the ladies prefer them also.

A classifier, or "grizzly," is a screening device for separating larger rocks from smaller gravel and sand to simplify the standard panning procedure. Any type of cover with holes one-quarter inch or so in size can be placed over the gold pan to permit small gravel to flow through with larger rocks examined and discarded. The "grizzly" can be a useful, time-saving device but should not be considered necessary.

GOLD PANNING KITS

Pre-packaged kits contain one or more pans and other screening devices sold as a unit with savings usually realized through a combination price. Quality must not be sacrificed, however, to save a few dollars. Your first gold pan should last many years and should be carefully selected. Beware odd colors and sizes. Choose a circular design and time-proven color. Additionally, the classifier must be designed for use on various sized containers and for use in either dry or wet panning.

Other than cost, the primary advantage of a kit is its ability to pack various items together compactly.

Many gold-panning kits include unnecessary items that can slow down the learning process and complicate your hobby. Avoid equipment using complicated procedures that vary from standard panning practices. There is no hand-operated device that will surpass field operation of a circular gold pan with built-in riffle traps. The bowl design has survived the test of time simply because it is the best.

NOTE: Of course, I recommend the "Gravity Trap" gold pan with its sharp 90-degree riffles, which I designed, field-tested and patented (U.S. Patent #4,162,969). These pans are manufactured and sold worldwide by Garrett Electronics, Inc. of Garland, TX.

Gold-panning kits can be a joy to use. They can save time and minimize your efforts, but they must be considered an optional item. Your primary purchase must be a standard-sized gold pan of the correct color.

The Garrett "Gravity Trap" Gold Panning Kit includes a classifier, two gold pans 14 and 10½ inches in diameter, a suction squeeze bottle to recover gold and a prospecting booklet. Pans feature 90-degree riffles.

GOLD SUCTION BOTTLE

The gold suction bottle, or "sniffer," is a small thin-walled plastic "squeeze-type" container. Inserting a plastic tube tightly into the bottle's mouth creates suction when the container is squeezed, and fine gold can be vacuumed through the tube and into the bottle. With the tube extending halfway down the bottle, gold is not forced out when suction is applied again. This procedure is carried out under water and is effective in removing fine gold from pans. Gold can be poured out of the suction bottle into a small vial. Although suction bottles are used extensively by experienced prospectors, they should be considered optional for the hobbyist.

GOLD TWEEZERS

Small, non-magnetic tweezers are useful for picking pieces of gold from black sand concentrate during the panning procedure. It is difficult to use fingers to remove these small gold flakes, especially while the panning procedure is continuing. You could lose sight of the gold flakes. If the suction bottle is used at this stage, you will pick up sand and other debris along with your gold.

MAGNIFYING GLASS

A small pocket-sized magnifier is usually needed to locate fine flour gold present in black sands. It is used in hard-rock prospecting to identify various metals and minerals in ore samples. A magnifier makes it easier to identify minerals present in the powder of a crushed specimen.

SMALL CAN

Any type metal container is satisfactory for bringing samples of crushed ore home for better classification. This procedure minimizes the need for working under difficult field conditions.

ORE BAG

Any type of sturdy cloth container is satistactory. Some hobbyists use specially manufactured "ore bags," but you can tie knots at both ends of a pants leg or shirt sleeve or purchase a "shot bag" at a sporting goods store. A bag is handy for carrying small ore samples in your pocket.

CREVICING TOOL

A crevicing tool is a steel rod, screwdriver or other long probe that can be used to force open a small crack or crevice. Gold tends to pack into small cracks in the bedrock, and a crevicing tool can be scraped through a crack to dislodge hard-packed gravel. Bending the point of the tool into a slight hook makes it even more useful. Crevicing tools can be purchased, but one can easily be made from any type of small steel rod.

ROCK HAMMER

A small hammer or pick is a favorite tool of most prospectors, especially for investigating hard-rock mineral deposits. It is used to drive rock chisels into cracks and to break off specimens. It can be used to split or crush rock, and the pick can pry up small pieces of bedrock. A model with a metal handle is less likely to break.

SHOVEL

After the gold pan, the shovel is a prospector's most necessary tool. Its uses are endless. Although a small shovel will usually be satisfactory I recommend a size number one or number two model for practical, all-purpose use.

SMALL BAR

Cracks in bedrock are almost impossible to open without a heavy prying tool. A small bar can be driven into the crack with a hammer to pry open a fissure or dislodge larger rocks. A small bar is almost mandatory in hard-rock prospecting since it is used to obtain specimens from mines or other excavations. Estwing Company manufactures a lightweight bar with a 90-degree angle on one end and the other sharply pointed for driving or prying.

LARGE BAR

The large, heavy bar is one of the most valuable tools in underwater dredging operations. It is not affected by water resistance when force is required to open cracks under water or dislodge heavy boulders. Difficult to drive with a hammer because of its length, it can be dropped or driven by hand with great effect. A large bar is the best alternative to dynamite for dislodging large rocks.

METAL DETECTOR

Metal detectors are producing more pounds of gold per dollar spent than any other type of prospecting equipment used in the world today. In fact, more gold nuggets have probably been discovered in recent years with VLF detectors than were found in the early gold rushes. Because

The Garrett Grand Master Hunter, world's most intelligent metal detector, is microprocessor-controlled and performs all electronic prospecting tasks. Previous deep-seeking Master Hunters have logged millions of hours in the gold fields.

gold nuggets are generally found in highly mineralized areas with large quantities of magnetic iron, it was almost impossible to find nuggets with earlier-model detectors. New VLF detectors with advanced electronic circuitry that ignore iron in the ground have been used to recover tens of thousands of gold nuggets. Verifiable stories of fantastic discoveries abound, and weekend prospectors are featured in many of them. Beginners and professionals alike have used metal detectors successfully. A metal detector designed for prospecting is relatively simple to use, but success comes only when the search area contains nuggets to be found.

Advertising by various manufacturers may confuse the first-time buyer of a metal detector. Select one from a manufacturer with a reputation for quality instruments in the prospecting field. Hobby-type detectors such as those used to hunt coins often cannot be adapted to mineral identification. Detectors designed specifically for hunting gold can recover nuggets in stream beds, search for mine veins, seek nuggets in dredge tailings, locate new gold field discoveries and otherwise prove useful in all aspects of field prospecting. They are highly successful when the minerals you seek are conductive.

Since few manufacturers have experience in the prospecting field, you should choose your detector carefully. VLF instruments generally cost only a few hundred dollars and are comparable in cost to a small gold dredge. Complete information on VLF detector is contained in my coauthored books, "The Complete VLF-TR Metal Detector Handbook," published by Ram Publishing Company.

TWO-INCH DREDGE

Several well-known companies and numerous backyard inventors manufacture suction gold dredges, ranging in (intake) size from one and one-half inches to eight inches. A two-inch size is the most practical for weekend prospecting. It is easy for the beginner to operate and costs only a few hundred dollars. When a paying location is discovered, a larger model can be purchased. There is generally some market for a used model, regardless of your success.

Small dredges are constructed in two configurations, described as "power jet" or "suction nozzle." The suction nozzle cleans bedrock and operates in water depths even as shallow as one or two inches. The power jet cannot be used in shallow water because it loses its prime when the intake nozzle is lifted from the water. The power jet creates more suction and is preferable where water depth is sufficient. A few manufacturers have combined the two principles of operating in a single unit. This is usually accomplished by using the same dredge and motor but changing the vacuum system through addition of different intake hoses and nozzles. The dredge used in taping the "Weekend Prospecting" video was a combination model.

Large commercial dredges and drag line operators did not get all the gold at any location. Much gold-rich gravel is still intact at early-day placer operations where creeks and rivers could not be drained and bedrock dried completely. Such operations are producing quite well for the modern suction dredge operator.

On rough bedrock the large commercial units could not dig up rock sufficiently to recover heavier gold trapped in fissures. Today, loosely packed gravel can be found in many of these locations, relatively easy to test with a gold pan or dredge. A suction dredge remains the only recovery system capable of vacuuming deep cracks under water without losing the gold. Complete small dredge descriptions and operating instructions are contained in my book, "Gold Panning Is Easy."

In conclusion, the weekend prospector should view prospecting as an exciting hobby with a good potential for profit. Concerning all equipment covered in this chapter and particularly dredges, he should begin slowly and not commit to any large expense until proficient in its use.

Chapter 3

GOLD PANNING

Gold panning can be divided in two phases: "wet panning" where water is available and "dry panning" in arid areas.

Your tools will include a panning kit with pans and classifier, shovel, waders or boots, gold sniffer bottle, pry bar, crevicing tool, tweezers and containers in which to bring home your gold and concentrate. You will not need the gold suction bottle and waders for dry panning.

WET PANNING INSTRUCTIONS

1. Place classifier atop the gold pan and fill the classifier with gravel.

2. Submerge pan and classfier, holding with both hands, and use a firm, twisting motion to shake them. Small gravel, sand and gold will pass through the classifier and settle in the gold pan.

3. Check for nuggets in the classifier; watch for clay balls that could contain gold.

4. Remove the classifier and discard its contents.

Step 1

Step 2

Step 3

Step 4

5. Grasp the pan with both hands under water and continue the rotating motion, occasionally shaking the pan. Remove small rocks as they continually move to the top of the contents. Occasionally tip the pan forward to permit water to carry off lighter material. Eventually, a small quantity of concentrate will be left in the pan.

6. Transfer contents to smaller finishing pan if a large amount of black sand is present. The following instructions apply to both large and small pans.

7. Make certain that Gravity Trap riffles are always on the lower side as you rotate the pan under water. This brings all material across the traps. Develop your own method for shaking, i.e. side to side, back and forth or a circular motion. Movement of the contents under water will cause the heavier gold to settle through the gravel into the riffles where it is trapped. As the contents become more concentrated, the total amount will grow smaller. Smooth and gentle motions are now mandatory. Contnue to tip the pan occasionally so that water can carry off lighter material. When visual identification is possible, try to separate black sand from gold by a gentle swirling motion of the water, leaving the fine gold concentrated together.

8. Retrieve your gold. Use tweezers for all large pieces and the gold suction bottle to vacuum up fine gold from a small amount of water.

Discard remaining black sand or save for later milling and further classification at home.

As you follow these instructions, you will develop your own methods of panning. No unusual type of equipment such as square or oblong pans will help because use of the standard circular pan is natural and easy for anyone to master.

You can practice at home by using small BB shot which will behave like little gold nuggets. More specific wet panning instructions are contained in my book, "Gold Panning Is Easy."

Step 5

Step 6

Step 7

Step 8

MOSSING

As water levels rise and fall, gold can be trapped in moss, grass, bushes and other vegetation. These plant systems form natural sluice box traps that can be worked year after year after waters rise and ebb.

To "moss" a creek or river, select one in a well known gold-producing district. As you walk along the bank, look for moss growing on tree trunks, exposed roots and boulders. You should also look for grass or other vegetation growing below the high water mark. Place moss or other vegetation in your large gold pan, taking care not to shake loose the dirt. Fill the pan with water and begin thoroughly washing the vegetation. After washing out all dirt and fine gold, discard the moss and grass and pan the contents.

Most gold contained in these natural traps is very fine and contains oil produced by the vegetation. Because the gold is both very fine and oily, you must be careful not to let it float off during panning. If this occurs, stop panning and dump the concentrate into your can for later separation at home. A few drops of Basic H (a Shaklee product), bleach or detergent in the pan of water will wash oil off the gold, allowing the metal to sink rather than float. You can then continue panning or use a gold wheel or bowl which are discussed in a later chapter.

CRACKS/CREVICES

Bedrock crevices are the most productive of all panning locations, especially for the weekend prospector. Crevices are traps created by nature and usually contain heavier concentrates deposited over countless thousands of years. Locating and cleaning such cracks and crevices is called "pot holing" or "sniping."

Gold is not found in all cracks, even in gold-rich areas. Gold may be absent because water flowed too fast in that particular section, the bedrock was too smooth or because of a combination of factors. The prospector should search for sections where bedrock is intersected by a natural iron dike or fissure, or is especially rough. It is desirable to learn to visualize where the channel was located in earlier centuries when the area was developing geologically. Older channels in which no water is flowing today can often produce significant yields.

The beginner should always stick to known mineral-producing areas for the most successful results. In such areas search banks of streams, rivers or higher ground where bedrock is sometimes exposed. A slick or worn surface on exposed bedrock will generally indicate a barren area caused by strong currents. If no bedrock is exposed, it will be necessary to uncover it by shoveling gravel.

Use the crevicing tool to rake or dislodge gravel from bedrock cracks. With the classifier atop the gold pan place gravel from the cracks into the classifier and move to the water for panning. Hold the pan and classifier together firmly under water and shake with a firm, rotating motion. Smaller gravel and sand will wash into the pan. Remove the classifier and examine its contents carefully for nuggets before discarding. Begin panning the remaining contents. If gold is extremely fine or there is a large amount of black sand, a smaller finishing pan can be used. Because the smaller pan can be handled more easily, only gentle motions are necessary.

The heavy bar is useful for crevicing or bursting bedrock, allowing you to succeed where others have failed. Sections of exposed bedrock may have been worked many times by inexperienced snipers who failed to get to the bottom of the crack where gold is generally concentrated. A small whisk broom is also helpful for gathering fine gravel and sand that is dislodged from such cracks. Sweep the rocks and cracks very thoroughly. Because gold is heavier than sand and tends to "stick" to dry rocks, it will remain in the smallest of cracks.

GRAVEL BARS/SANDY BEACHES

Sandy beaches can be pitfalls for novices who attempt to pan or use a suction dredge on them. Except for widely known Gold Beach in Oregon and Nome Beach in Alaska, sandy beaches rarely produce anything but microscopic gold. Presence of glittering mica in sand also confuses beginners since it resembles fine gold. The experienced weekend prospector has no trouble identifying mica, but the novice must continually be reminded that all that glitters is not gold. Bypass sandy beaches, especially while learning.

Large gravel bars with small-to-medium-sized gravel can be found in most any river or stream. Such bars are generally alluvial deposits of recent origin and are loosely packed. Little of value other than microscopic fine gold is usually found in loosely packed gravel. Even large commercial dredging units are not always successful in processing alluvial gravel bars because such large amounts of gravel must be processed for only a little gold. Bypass loosely packed gravel bars.

Search for tightly packed gravel or gravel in areas that might have been missed by early prospectors. Look too for signs of oxidation that indicate the gravel is old or was part of the original stream bed. Practice will make it easier for you to find such spots. They may be under large tree roots extending into the stream bed, under overhanging banks or even high on a hillside (a "high bar" deposit).

In areas that were dredged by early-day placer operations try to locate spots where all of the original deposit was not taken out or worked. Shovel gravel into the classifier atop the pan and place under water. Proceed with standard panning procedures. In some areas you can reach into the water with a long-handled shovel to recover enough gravel for a good test. Locations rich enough to make this method of sniping practical can be found behind large boulders and in deep, slow-moving water where currents deposited the heavier gold. Testing for values by using the gold pan is relatively fast and the results are often favorable.

DRY PANNING

Dry panning is not the most reliable method of discovering gold in desert areas, but it is often the only practical one. Sharp and deep riffle traps in a pan are required. These should be positioned at a 90-degree angle to the wall of the pan. Gravity Trap riffles are designed like a small dry-washing sluice box. If operated correctly, they produce good results, even where water is not available.

Likely locations in the desert are the bottoms of gullies or dry washes. Use basic tools for bursting bedrock and removing gravel.

DRY PANNING INSTRUCTIONS

1. Place the dirt in a large Gravity Trap gold pan.

2. Clean the larger rocks by rubbing them together or striking them against each other sharply. Let dirt and gravel fall into the gold pan.

3. Discard thoroughly cleaned rocks.

4. Sift your fingers through the small rocks remaining and lift them out without removing any fine gravel or sand. Examine these rocks closely for possible nuggets before discarding. Continue until all that remains is fine concentrate.

5. Lift the pan in one hand and tilt at a 45-degree angle with riffles held on the lower side. This permits contents to flow over the riffles, and a correct panning procedure "traps" gold in the riffles. Use your other hand to bump the high edge of the pan. Sharp bumps will vibrate the contents as it flows over the riffles. Continue bumping the pan and occasionally level it to shift contents to the bottom. Repeat this procedure as lighter material gradually flows out of the pan. Finally only one or two handfuls of concentrates will remain.

6. Inspect the concentrates by spreading them over the pan. Blow sand away. You will occasionally expose a small nugget or gold flake.

7. Retrieve nuggets or flakes with tweezers.

8. Save remaining concentrates for later wet panning in camp or at home.

If you are evaluating several areas, carefully label each

concentrate sample. Many areas rich in gold have been discovered by these dry panning "sniping" procedures. They were then worked later with conventional dry washers. Often rich deposits from ancient rivers are found high on mountainsides. Effective dry panning techniques minimize the sizes of samples from such locations that must be taken to camp for wet panning. This sometimes enables the prospector to pinpoint precisely his richest locations.

It is difficult to carry a dry washing unit into deep canyons and other remote locations to test for gold. Because dry panning with a Gravity Trap pan is easily accomplished, such a gold pan should always be carried even when prospecting in areas remote from water.

Gravel tested in a dry washing unit must be thoroughly dry. Even a small amount of moisture will slow or halt the settling action of gold in a dry washing unit. Dry panning can, however, concentrate moist gravel to a reasonable degree.

It is true that dry panning is far from a sure fire method for getting rich by finding gold. Use of this technique, however, can make it unnecessary to pack hundreds of pounds of material samples over many miles or desert or mountains. Dry panning can also prevent packing a heavy dry washing unit into a remote area before gold values in the area are tested thoroughly.

Watch for areas far from water with gravel. Some of the richest deposits have been made in such locations. All of these certainly were not discovered by dry panning, but proper use of the technique could have determined their values.

The desert contains many fortunes in gold that have been overlooked. Some were never tested simply because water was not available. Remember dry panning whenver you pass a likely spot only where water is not handy. If in doubt, use your Gravity Trap gold pan. More specific dry panning instructions are contained in "Gold Panning Is Easy."

Chapter 4

GOLD DREDGING

A surface suction dredge is the best equipment for successfully recovering gold from beneath water. A two-inch size is probably best for the weekend prospector because it is easy to transport and simple for one person to operate. The suction nozzle dredge is the ideal choice for prospecting in shallow water. A power jet model is better suited for depths of six inches or more. Models incorporating both suction and power jet features cost only a little more than a unit with only one function.

The weekend prospector may also need these items:

—Gold panning kit
—Gold sniffer bottle
—Waders or rubber boots
—Gold tweezers
—Inexpensive face mask
—Magnifying glass
—Small can or other container
—Crevicing tool
—Large or small bar
—Metal detector to aid in location of black sand

GOLD DREDGING INSTRUCTIONS

1. Fill the dredge with gas and oil.

2. Position the dredge on location and place the foot valve and suction hose under water. Make certain the nozzle end is competely submerged. Manufacturer's instructions should be followed carefully since some dredges are self-priming, and others must be hand-filled.

3. Start the motor (choke as necessary). If the pump does not start pumping water IMMEDIATELY, stop the motor to prevent damage to an unprimed pump. Repeat the priming procedure and start the motor again. The suc-

tion hose should create a good vacuum with water flowing freely over the sluice box. Power jet nozzles must be kept under water at all times, or else the pump will lose its prime and must be re-primed.

4. Begin dredging operations. With one hand grasp the nozzle and place one or two fingers across its opening. Let gravel gradually feed into it. Dig gravel loose with your other hand. Remove large rocks to keep the operation running smoothly.

Jamming the nozzle into loose gravel will cause it to plug up. If this happens, the motor must be stopped and the hose cleaned of rock and sand. Restarting procedures may create surges of water that disturb the sluice box contents. Strive to keep a uniform flow of water over the sluice box. This can be achieved after only a little practice.

5. After the sluice box has filled to a satisfactory level, empty the contents into a gold pan or large bucket for later

Gold pans in hand, Virgil Hutton carries a backpacker dredge to a wilderness site. Manufactured by Garrett Electronics, the two-inch dredge is lightweight, extremely portable and can be adapted with both suction and power jet nozzles.

panning. Simple and easy for one person to operate, a surface suction dredge is the only practical method for removing gold values and nuggets from underwater cracks and crevices. Detailed instructions and more specific operating information are contained in my book, "Gold Panning Is Easy."

SHALLOW WATER CREEKS

A small suction dredge operating in only a few inches of water will "work" a small amount of gravel compared with the capability of larger models. A small dredge expertly operated, however, can produce more gold than a larger dredge in the hands of a novice.

Site selection will ususally determine results. Many early day placer operations were not suitable for sluice box operation, or the bedrock itself lay so far below the water table that it was impossible to recover all gold efficiently. Look for spots where these conditions might have occurred.

Bedrock of small creeks can often be found at a very shallow depth, and removing only a foot or so of overburden (gravel and dirt) permits inspection. Let me emphasize that gold values are not concentrated solely on bedrock. Generally speaking, however, bedrock is a sure gold trap and a profitable place to begin operations.

Shoveling away the overburden is quicker than removing it with a dredge. Often, the water will be shallow enough for you to feel the bottom and locate cracks and crevices. Use a small bar or crevicing tool to clean out cracks and loosen all hard-packed gravel. A large bar should be used to burst the bedrock to determine if there was a "false" layer which is indicated by a "rotten" or "loose shift" formation when bedrock is located. Then use the dredge to vacuum loose material and pan contents of the sluice box for further separation.

When cracks in the bedrock are visible, use a large bar to open them so they can be cleaned out thoroughly with

This suction nozzle is suitable for dredging in shallow water because it will not lose prime when lifted from the water. Note small hose ejecting water from dredge pump into U-shaped nozzle near the end of larger bent nozzle to create suction.

a crevicing tool. Vacuum the loose material and inspect it. You may follow a small creek for several miles and obtain little gold or it may be barren. Do not become discouraged; continue testing. Often the creek has changed course or water currents are different from those when gold was deposited. Look for areas where a high water table might have hindered earlier prospectors.

Boulders in the water table are promising. Vacuum on the downstream side or underneath them. Dredging can dislodge such boulders, however, and you should avoid being trapped by them. This is a common accident for beginners and is sometimes fatal.

Inspect feeder streams even if they carry only a trickle of water. These drain from higher elevations that could contain rich gold deposits, and they may provide the source of gold in larger downstream creeks. Use your shovel to create a hole large enough for dredge operations. Build a dam to create a reservoir of water for recycling.

In arid desert locations a small dredge is often helpful during spring runoff conditions. It offers a much faster method for testing values than dry panning. Arid districts with only occasional flowing water often have different types of deposits. Procedures for testing them may be different in comparison with those used in areas consistently under water. Clay material particularly requires careful testing.

Select any small wash or ravine during spring runoff and dig a hole to remove some of the overburden. Weekend prospectors must proceed as quickly as possible and use the easiest method with available tools. Arid areas are often composed of small shale, decomposed rock or granite which is easy for a small dredge process. The challenge is to discover enough water for dredge operation.

Beware of deceptive areas that may look safe during dry periods but become flood torrents during rainstorms. Prospectors have lost their lives working a dry wash or ravine.

SANDY BEACHES/GRAVEL BARS

We have already noted that areas such as sandy beaches and loosely packed gravel bars are usually unproductive and unprofitable for the small dredge operator. Loose gravel seldom contains little more than fine gold values, and sandy beaches often have none at all. The limited production capability of the small dredge makes such areas even more unprofitable. Gravel bars can be worked easily, however, an activity that can prove interesting if profit is not your primary motive.

On the other hand, sandy beaches usually produce only a worn out water pump since bearings on all high-pressure pumps are adversely affected by sand which will get in the foot valve. Also, light sand or fine gravel must always be fed slowly into the suction nozzle, or the hose will become clogged. A dredge operates like a vacuum cleaner; when its suction hose is overloaded, it jams.

If a weekend prospector wishes to test areas with sand or fine gravel, the foot valve should be placed under water inside a small bucket to keep sand from being picked up. The operator must watch closely for jamming conditions while slowly feeding the suction nozzle. These guidelines apply anywhere sand or small loose material is encountered.

DEEP WATER RIVERS

Dredging in deep water is limited to shallower spots where the operator can gain access to the bottom. Occasionally it is possible to work in waist-deep water, but even then the operator must put his head under water from time to time. Of course, this is not practical for extended periods but the procedure permits testing areas with deep water. An inexpensive face mask and snorkel allow the operator to observe his work under water, provided the water is sufficiently clear.

Cracks and crevices remain primary targets, except that some bedrock cracks in deep rivers are wide and deep. Even when the bottom of these cracks can be reached with the nozzle, its size may prohibit the nozzle from reaching bottom where most gold is found.

A smaller nozzle of flattened metal tubing rarely works because it becomes clogged too easily. I suggest carrying a piece of short hose which can be inserted into your suction nozzle. Collapse it slightly, and it will permit sufficient penetration of small cracks. When it becomes jammed, you can clear it easily by allowing it to expand. This method enables you to vacuum small gravel and gold from deep and narrow crevices.

Large boulders create a natural trap for gold that moves constantly through flowing water. Because annual high water replenishes the supply, most gold remains near the surface of the gravel in this type deposit. Such deposits are among the easiest to test in deep water because there is no overburden. Simply vacuum the loose material on the downstream side and vacuum under the boulder as much as possible. If the water is too deep, use a longer nozzle or pipe with your suction hose with the end crimped like the original nozzle. It will become jammed frequently, but a screwdriver or other rod can be used to dislodge jams with the nozzle held just submerged to avoid losing the prime on the suction hose.

Some weekend prospectors combine pleasure and profit while "floating" rivers. When large boulders or other natural gold traps are spotted, they simply stop their float, start the dredge and vacuum loose concentrate trapped on the downstream side. Such prospecting for alluvial gold deposits is one of the easiest and most profitable ventures for a small dredge owner and can be enjoyed by the entire family with little physical labor.

Virgil Hutton, Tommy Long and Charles Garrett "snipe" along the Snake River in Idaho, searching for black sand concentrations and placer. Garrett Master Hunter detector locates deposits that can be cleaned out with a dredge.

Each year water currents deposit new concentrates for the small dredge operator to find. You should examine areas where larger dredges have worked. The large size of their suction nozzles often forces some operators to leave gravel at the bottom of deeper crevices. When you spot remaining hard-packed gravel, use your crevicing tool to loosen it and your smaller nozzle to clean the bottom.

The previous dredge operator cleaned bottom so you could "snipe" his operation with your smaller dredge. These deep cracks usually contain the heavier gold and can contain more than 75% of the total recoverable gold values. Always save black sand discards from your dredging sessions for later cleanup and profitable separation at home.

Chapter 5

NUGGET HUNTING

The term "nugget hunting" is so ambiguous that no description of it could ever be complete. Countless books have been written about this method of searching. Weekend prospectors generally find their instructions too complicated. By condensing descriptions of target areas and summarizing the easiest searching methods for beginners to use, I offer you instructions that are simple but that employ methods proven successful the world over.

Tools you may include:

- Gold pan or gold panning kit
- Waders
- Gold tweezers
- Rock hammer
- Small bar
- Shovel
- Metal detector (suitable for use in mineralized areas)

The metal detector will unquestionably produce the best results in areas where gold abounds and large nuggets commonly occur such as the western United States deserts, certain "mother lode" areas, Australia, New Zealand, China, Africa and other parts of the world where sizeable nuggets appear in nature. Fantastic finds with metal detectors are regularly reported.

NUGGET HUNTING INSTRUCTIONS

An electronic metal detector can be used effectively because gold is conductive and its presence will be indicated.

Unfortunately, so-called "hot rocks" contain an iron mineral content that is either greater than or less than the levels for which the ground-cancelling detector has been adjusted. They present a constant nuisance to prospectors because they register as metallic (conductive) on a detector. Calibrated circuits used in modern metal detectors that allow easy identification and rejection of "hot rocks" are an absolute necessity in helping a prospector determine whether a rock contains gold or a predominance of iron. Make certain that your detector is a ground cancelling type that contains these special circuits.

In wet areas start your search near water.

1. Adjust your detector to the magnetic iron content of the area according to manufacturer's instructions.

2. Pinpoint your target.

3. Slip a shovel under the target and place it into the plastic gold pan. If it is located in a bedrock formation use your rock hammer and small bar to dislodge it. Then place the dislodged target into the pan for examination.

4. Check the contents further with a detector. If a "hot rock" is suspected, use the detector to determine if it is ore. If not, discard it.

5. When the detector indicates gold is present, inspect the contents visually or follow panning procedures.

6. When the detector indicates no gold in the pan, repeat Steps 3-5 until the target is located.

Metal detectors used for prospecting will also indicate large concentrations of black magnetic sand. Shovel the sand into your gold pan. Inspect for gold nuggets and save the black sand discards for later separation.

In dry areas the procedure will vary only slightly. Locate the target with your detector, dig carefully with hands or a small tool, being careful not to damage the nugget.

Metal detector earphones are an advantage in most areas since small nuggets generate only a faint response. It is best to dig or investigate visually all targets unless they can be identified absolutely as "hot rocks." Electronic discrimination is a valuable aid but should not be relied upon entirely.

Areas with small, loose material make visual identification of targets more difficult. When searching such areas, shovel targets into a plastic gold pan or small plastic cup and check for electronic responses. When your detector indicates that a metal object is in the pan, use dry panning first to reduce contents. Then grasp a handful of the pan's remaining contents with your hand (which must be free of rings and other metallic jewelry) and pass your hand across the searchcoil's detection area. If you use a plastic cup, it can be inspected in the same manner. Make certain your detector is tuned correctly and move your hand containing the material across its coil. Continue until you get a response, then place contents that generated the response into another pan to avoid loss of the target.

Cracks and other bedrock sections where gold may be trapped should be inspected in a similar manner. More detailed instructions are contained in my co-authored book, "Electronic Prospecting," published by Ram Publishing Company.

In desert areas where medium to large nuggets occur and water for testing them is scarce, the metal detector provides the easiest method of recovery. The introduction of VLF metal detectors has brought with it fantastic success stories. Natural elements continually erode mountains, allowing rich deposits to surface. Once a gold nugget of sufficient size becomes exposed, it can be discovered by a metal detector. These nuggets are rarely detectable by sight alone and the absence of water leaves electronic detection as the surest and most effective method.

In many areas of the western United States gold is too fine for electronic detection, and dry washing machines or dry panning must be employed for recovery. Almost all areas where gold is now located once contained an original vein or concentration of ore that weathered over the ages into a placer deposit. This means that nuggets may still be present even in areas where the gold has been too fine to register on metal detectors. Presence of such valuable nuggets is continually proved as successful searches are conducted with detectors in the dry washes, arroyos and canyons of arid locations.

Streams can be a valuable source of nuggets. In heavily mineralized areas where productive mines are located, rich ore specimens are often deposited in streams by natural erosion. All targets should be carefully examined before assuming that one is merely a "hot rock." Valuable coins can often be found in streams of old mining districts. The silver-producing areas of Mexico also produce large nuggets that can be easily recovered from creeks and rivers. Small streams created by the melting of large glaciers in Alaska and western Canada often contain nuggets easily found with detectors.

Tommy Long uses a Garrett Master Hunter detector to search for nuggets in a stream bed. He is using a four-inch Super Sniper searchcoil with both extreme sensitivity needed to locate tiny nuggets and maneuverability to search a rocky area properly.

Searching under water for nuggets with a metal detector is often not as profitable as in other areas. Considering the small amount of labor required, however, especially in comparison with dredging or panning, underwater exploration is well worth the time and effort. Search slowly and carefully. Examine all target responses; usually metallic junk is present in streams. And, "hot rocks" are present in water just as on land. Very often, companions of dredge operators use their spare time to search streams for nuggets. Persistence and patience are the keys to success here.

Large nuggets encrusted with a black or dark coating have been found particularly on mountain tops. It is believed that volcanic action or oxidation of other minerals and materials encrusted gold with the black coating. Such discoveries commonly called "volcanic gold" or "black nuggets" represent a rare opportunity for the prospector and are almost impossible to locate except by electronic detection.

Literally millions of dollars in gold nuggets are being discovered all over the world. Metal detectors enable suspect areas to be searched in a manner never before possible. It is reasonable to state that more nuggets have already been discovered with metal detectors than were ever discovered in all the old gold rushes. The use of a metal detector will provide the weekend prospector with many enjoyable and exciting hours of recreation and can detect riches beyond anyone's wildest imagination.

Chapter 6

FIELD SEARCHING

The term "field searching" could easily cover the entire spectrum of prospecting. Because the weekend prospector seeks results with a minimum of effort, we will limit our discussion to areas and conditions most commonly occurring and producing positive results. Remember that volumes have been written concerning most major geographic areas where large discoveries have been made. Technical manuals detailing these discoveries, however, are written by professional prospectors and others in the mining field. The following suggestions are quite simple in comparison, yet they include procedures used by most successful prospectors today.

To get started you will need:

—Gold pan or gold panning kit
—Gold tweezers
—Magnifying glass
—Ore bag
—Rock hammer
—Small bar
—Shovel
—Metal detector suitable for prospecting

The metal detector should have an assortment of searchcoils. Small searchcoils permit detection of veins near the surface, while larger ones can detect small pockets of conductive ore.

FIELD SEARCHING INSTRUCTIONS

1. Attach your largest searchcoil to the metal detector.

2. Tune or adjust the detector according to local conditions and the manufacturer's instructions.

3. Search the area. With the detector searchcoil held directly before you, not sweeping, walk across the search

area. Establish grid patterns approximately 10 to 20 feet apart.

4. Investigate all target responses. Because iron deposits often contain valuable minerals, both gold and iron ("hot rock") indications should be investigated. Discrimination features on quality detectors help you determine if an indication is a non-ferrous, highly conductive vein of ore or of a magnetic iron dike that is only faintly conductive. It can also indicate whether an iron deposit contains other minerals of interest to the prospector.

Detection depth is usually limited to no more than 20 feet and can be much less, depending on how close to the surface the so-called "halo effect" will produce a signal. A small amount of digging will often expose detected veins, but sometimes earth-moving equipment is required. When the vein is exposed, use your rock hammer and/or small bar to retrieve ore specimens for evaluation at home.

Ray Lagal field searches with VLF detector, using an eight-inch diameter searchcoil for close examination of exposed ore veins. This general purpose searchcoil also detects small nuggets and identifies ore specimens.

5. Attach a small searchcoil and make certain the detector is tuned properly. The smaller coils are better for close examination of exposed ore veins, detection of small nuggets and correct field identification of ore specimens.

6. Using the smaller coil, sweep the detector over veins you have uncovered or other deposits that interest you.

7. Store samples in your ore bag. If the pocket material is loose or if secondary enrichment is suspected, later inspection by wet panning procedures or other more sophisticated methods will be necessary.

Complete instructions regarding search procedures are given in my book, "Detector Owner's Field Manual."

Logging roads, trails, highway cuts or anywhere man has used machinery to cut into hillsides or to expose areas beneath the earth's surface present a field searching opportunity for the weekend prospector. Because prospecting was not the intent of those who made the cuts in the earth, these locations have probably never been inspected. Construction crews are seldom allowed time for prospecting.

Road-building and other activities requiring earth to be moved can expose ore veins, clay deposits containing rare earth or mineral values and even ancient river channels. All of those could contain gold nuggets.

Always be alert to the presence of drifts anywhere you travel, since this can indicate an old river channel or glacial deposit. Such gravel can be found anywhere—high on a mountain pass or on a desert plain. Take a minute to secure and mark specimens for evaluating at home.

Drifts are often indicated by the oxidation of ferrous minerals that result in discoloration or signs similar to rust. Such veins discovered on roads and trails have provided profitable "one-man" operations for many prospectors. When you encounter such rust signs, use your hammer and smaller bar to retrieve samples for testing.

On rivers where early-day placer operations were conducted, watch for small tunnels in the hillsides. They generally were dug to follow some fault or bedrock crevice suspected of having trapped gold. Let me emphasize that these tunnels were dangerous when they were dug through loose gravel years ago. Commercial operators continually discover old tunnels containing the bodies of early-day placer miners. The tunnels have grown doubly dangerous with age, and you should never attempt to enter them when alone. It is best that you be accompanied by a prospector who is experienced in dealing with such tunnels.

If you become convinced that a safe inspection of some tunnel is possible, however, you might find gold nuggets with very little effort. Remember that days or months were required to drift the tunnel along bedrock to inspect only a few yards of material. You can easily inspect the sidewalls and lower edges of the exposed tunnel without removing any material at all. Using your detector to probe electronically into the gravel behind walls and floors often

Roy Lagal, Tommy Long and Charles Garrett investigate a mineral dike while field searching with a VLF detector equipped with Garrett's Depth Multiplier attachment which multiplies depth capability of the detector up to four times.

enables you to examine more cubic yards of material in just a few minutes than was examined by miners who might have taken months to dig the tunnels.

Always be alert to danger when dealing with tunnels When your detector indicates a target, extreme care must be used in removing it so that you do not dislodge the roof of the tunnel or otherwise cause a cave-in. Sometimes just a slight noise or other vibration can result in disaster.

Earthquakes, storms or other acts of nature often create cracks in the earth's surface or displace its formation in other ways. Mud slides will expose an entire mountainside. Such cataclysmic events present new areas of exploration for the alert prospector requiring little effort. Valuable mineral veins and other deposits have been uncovered by natural disasters. Of course, caution is always necessary when working in the area of a natural disaster, especially if you descend into exposed cracks.

Be alert for the scenes of natural disasters when traveling through heavily mineralized areas. Many times minor rockslides, slight earth movements and other occurrences will go unnoticed by local residents. Use both visual and electronic methods to inspect any unusual indications and obtain samples for more sophisticated testing.

OPEN FACE MINING

Large veins of quartz often appear on hillsides or on the earth's surface, and these are usually barren of any mineral values. When you encounter one with signs of oxidation, however, it should be tested electronically for metallic indications. If your detection signals prove interesting, remove a specimen, use a magnifying glass to inspect it visually and save for later evaluation.

In field searching, the successful weekend prospector takes full advantage of the assistance provided by both man and nature in years and centuries past while utilizing the most modern electronic methods available to him. If you will seek out these opportunities to discover and identify precious minerals, you can cram an incredible amount of prospecting into one short weekend. To further enhance your enjoyment of the hobby, learn more about the areas you are examining through research. This can help you avoid wasting time in fruitless prospecting.

Chapter 7

MINE SEARCHING

Old drifts, tunnels and other areas of underground exploration can offer a bonanza to the electronic prospector. This is not meant to exclude newer mines, but the newer locations are likelier to have encountered electronic detection. In older mines abandoned years ago, however, all the hard work of earth moving has already been completed, and it is possible that the operator missed a vein or "pocket" by inches.

Often a rich piece of high grade ore was left just beneath the surface. If the deposit is of a conductive substance, the metal detector will indicate it. Your detector can quickly prospect any type of mine in little time with hardly any labor required.

In fact, even if you only examine walls, floor and roof of the mine a few inches deep you will have prospected more cubic yardage than the original miners who moved tons of earth and rock.

Before continuing, we must warn the weekend prospector of the dangers always present in mines both old and new. They should NEVER be entered alone. Tunnels, drifts and shafts often contain deadly gases, unstable explosives, rotted shoring and fractured rock structures. Each of these is just waiting for a spark or slight vibration to cause a disaster. You should always be accompanied by an experienced professional when dealing with mines.

MINE SEARCHING INSTRUCTIONS

Select a metal detector capable of cancelling iron minerals. Your detector must also be able to identify ferrous and non-ferrous deposits and measure their conductivity.

1. Attach a large searchcoil to the detector for maximum depth penetration.

2. Adjust the detector for local conditions in accordance with manufacturer's instructions.

3. Search the area with a slow scanning motion of the searchcoil. Make sure the detector remains in the same position relative to your body to prevent false signals coming from your light, shovel or other tools and metallic items. The searchcoil should normally be six inches to one foot away from the search area, but this distance can be altered in relation to the presence of magnetic iron. False signals can be caused by uneven surface areas. Pay close attention to the roof because an exposed piece of rich ore could have been overlooked. Also, carefully scan the floor. If a mine was producing high-grade ore, the floor will always contain a few small but rich specimins.

4. Investigate all targets. Often a deposit may be only slightly conductive but still be enormously rich. When a target is detected, use the detector's discrimination feature to determine its content. The target may be difficult to examine because of small pieces of iron, tools, blasting caps or other refuse present. If so, a specimen should be removed by hand and examined with the detector coil.

5. Always obtain specimens when in doubt. Often the target can be identified as worthless iron, but specimens should be obtained with the rock hammer or small bar for later evaluation.

In mining districts where rich gold pockets were common, secondary enrichments often occurred. This phenomenon was caused by the deposit's being leached or decomposed, with gold values becoming trapped far away from the original pocket. Such a fine deposit will usually cause only a very slight audio response; in fact, fine gold is often not conductive enough to cause any audible response on a metal detector.

Selecting old mines at random for exploration seldom produces profit. Research the area by asking questions and reading reports. Obtain milling information, if available. Determine if the mine or area produced high-grade ore and whether it occurred in some complex ore or was a free-milling type. Generally, the free-milling type could be seen with the naked eye and was in a vein or pocket formation. It is usually highly conductive and easily found with a metal detector. Gold in a complex ore could be very rich but not in a conductive form and, thus, difficult to find with a metal detector. These statements apply to any precious metal.

Finally, local ore specimens are often displayed in nearby museums. Viewing them can help you determine whether the area offers opportunities for electronic prospecting.

One advantage to exploring older mines comes from the generally ineffective milling techniques employed in early years. Gold or silver had to be in a native state for extraction to be possible. Therefore, values in older mines are almost always in a metallic state and highly conductive.

In Mexico literally tons of pure native silver has been recovered from old Spanish mines with the aid of metal detectors. The almost pure silver was unidentifiable with the naked eye because it was slightly covered. Unfortunately, access and resale possibilities in Mexico present great difficulties, making electronic detection here a moot point.

Arizona mountain ranges are rich in native silver and produce good results for those whose research helps them locate conductive ore patterns. Canadian silver mines in the Cobalt district offer unlimited opportunities to the electronic prospector. The silver is almost pure and in a native, conductive state. The only remaining obstacle is obtaining permission to inspect the abandoned sites.

Shallow gold mines in Australia and New Zealand are producing fantastic finds when reworked with metal detectors.

Once again, the weekend prospector must always bear in mind that all this exploration will take place underground and presents extreme danger. If exploration can be safely accomplished and the values are in a conductive state, rewards could make you rich.

Old mines are also covered in my book, "Detector Owners Field Manual."

Chapter 8

MINE DUMPS

The term "ore dump" generally refers to the location where richest ore from a mine or mines was stored prior to milling or transportation from the site. You must first determine which was the ore dump and which was the "tailing pile," containing generally valueless mine waste. You will usually find valuable minerals in the area where rich ore or vein materials were stored. Remember, however, that human error or thievery could have diverted valuable ore into the worthless tailings pile. Such diverted material was later recovered by unscrupulous employees engaged in what was commonly called "high grading." It could also be called stealing.

Other than desert areas where electronic prospectors are achieving amazing results locating large gold nuggets, the most profitable areas for electronic prospecting are the old ore dumps where rich specimens might have been overlooked for one reason or another. Someone else has completed the work of bringing this material from beneath the earth; your chore is but to analyze it electronically.

ORE DUMP INSPECTION INSTRUCTIONS

1. Use a metal detector with ground cancelling and added discrimination. Attach a medium or large searchcoil.

2. Adjust the detector to local conditions according to manufacturer's instructions.

3. Search the area. First, scan the material at a height of four inches to one foot, depending on the amount of interference encountered from any magnetic iron.

4. Investigate target responses. When targets are located, dig or chip them out and isolate them for further testing. Place the detector in a prone position and determine if the target is conductive or worthless "hot rock" by

using the discrimination mode on your detector. Accuracy is not always possible with on-site inspection, but it can be helpful, especially if you will familiarize yourself with different ores and metals known to be present in the area you are searching. Always test target samples for conductivity with the detector in a stationary position. Move speciments into and away from the searchcoil's center.

5. Place specimens in an ore bag and bring them home for better evaluation. "Hot rocks" will usually be rejected at the dump by using only a small amount of discrimination. When visual inspection indicates that a target is located in a larger worthless rock, use your rock hammer to break out the target. This will avoid transportation of worthless material.

Weekend prospectors are shown investigating mine dumps with VLF detectors that have been ground balanced precisely. They recommend the detector use no discrimination here or in dredge tailings to insure that all nuggets and ore are detected.

Many mine dumps have been declared "off limits" to casual prospectors by owners who have become aware of their value. Owners are now "leaching" their dumps with cyanide or using other methods to recover lost values. If you believe that an ore dump might contain values in the native or conductive state and searches are not permitted, try to strike a bargain with the owner by letting him share in your prospecting results.

You may be suprised how many mine owners are either unaware of electronic detection methods or have little confidence in its possibilities. The owner may agree to let you explore old dumps since it costs him nothing and might be profitable to you both.

Native silver will often be detected within worthless rock which can be burst to expose the almost pure

George Mroczkowski, treasure hunter and author, displays a silver specimen recovered with a metal detector from an ore dump (Chapter 8). Knowledge from this book and the proper detector enable weekend prospectors to be just as successful.

specimen. Gold nuggets can also be found in this manner. Small quantities are hard to spot visually and will require bursting larger pieces of rock for hand-testing of each small specimen with the detector placed in a prone position. Even faint responses can be readily interpreted.

Be alert for old dumps that are now covered by vegetation, especially in remote areas. Because identification of values in these older mines was often almost entirely visual, rich specimens may be discovered. We have learned that while the eye can only view the surface of a rock, a metal detector can look inside it.

In areas where small surface-type gold pockets were common the "pocket hunter" extracted ore generally from a small hole and left behind a mound of what was judged to be worthless rock. Because of the incredibly rich content of some of these small pockets, pieces of the discarded rock contain small gold stringers or nuggets. Normal searching will seldom locate these small but valuable gold specimens. It is best to place the detector in a prone position and use hand-test methods to evaluate them. Presence of conductive targets indicate the hole probably was a rich one, and you should continue searching. When no values are located after a few tests, move on.

In the United States, Canada and Mexico, where gold and silver often occurred in the almost pure, native state, electronic prospecting of old mine dumps is the most productive type of weekend prospecting. Your only challenge is to gain access to the dump of a mine—either old or new—that produced free-milling ore. Large specimens with almost pure metallic content and weighing several pounds are being recovered daily at dumps by prospectors using metal detectors. You can see many of these specimens on display at various locations around the world.

Chapter 9

DREDGE TAILINGS

Entire river valleys have been dredged by large operators. Before environmental legislation enforced restoration of the land, huge piles of tailings were usually left wherever they were discharged, creating landscape eyesores that are still visible. When research produces evidence that large gold nuggets occurred in a particuar mining district, old tailing piles of dredge operations can be a good location for electronic prospecting.

Large dredges generally ripped up gravel and overburden down to bedrock and deposited it in a "trommel," or classifier. Larger rocks, usually over one-half or one inch in diameter, were discharged by conveyor belt onto a tailings pile which can be easily recognized today. Smaller gravel, sand and gold values passed through the trommel into a sluice box where the gold was recovered.

Rocks discarded because they were too large for the trommel sometimes contained nuggets. Of course, dredge operators continually inspected the conveyor belt for nuggets. Often the inspector was absent or inattentive enough to permit nuggets to be discharged with tailings. At other times fine gold and nuggets were concealed in large clay or mud balls.

We emphasize that all dredge tailings will not contain nuggets. You can find them only in those mining districts where large nuggets occurred in nature. Some research will be necessary to prove the good locations. Generally, local residents will know if nuggets were discovered in an area.

DREDGE TAILINGS INSTRUCTIONS

1. Select a metal detector with discrimination mode that identifies "hot rocks" from gold nuggets and use a medium or small searchcoil.

2. Adjust the detector to local conditions according to manufacturer's instructions.

3. Search the pile. Often small pieces of junk iron will be found in dredge tailings along with "hot rocks," but indentification should never be attempted without closer inspection. Your target must be dug out of the pile. This is true of all nugget-hunting, regardless of location. Magnetic iron content of heavy mineralized areas presents too many possibilities for error.

4. Examine targets. In dredge tailings be very careful in removing rocks that cover a target; continually check for responses as you expose it. Sometimes just a small vibration will drop your target rock deeper into the pile. When a small target is located among gravel or rocks, slip a shovel under it and carefully place its contents in your gold pan for examination with the detector. Often, visual

This gold nugget worth thousands of dollars was found in old dredge tailings. Site along rivers dredged long ago have yielded nuggets weighing in pounds. Searching tailings and mine dumps is easy since the hard work of moving rocks has already been done.

identification may be possible. Continue this procedure until all your targets are located and either identified or packed away in an ore sack for further inspection at home.

Recovery of gold nuggets from old dredge tailings should be considered a profitable weekend prospecting endeavor. It is worthwhile because large nuggets can sometimes be found. The tailing piles themselves are generally easy to reach. Furthermore, the search usually becomes a pleasant outdoor activity. In remote districts dredged streams are usually a popular scenic spot for picnickers, hikers and others who enjoy the outdoors.

Use of a metal detector by a weekend prospector can go unrewarded for many days or weeks, but picnickers and semi-professionals have been successful at dredge tailings. Most outdoor enthusiasts are rockhounds at heart and cannot resist digging into any discarded rock piles, especially in areas where gold has been found. I once recovered a two-and-one-half-pound nugget from dredge tailings in one of the better known mining districts in the Northwest. Even visual inspection of dredging piles can result in the discovery of gold nuggets or other mineralized specimens.

I urge you to consider carefully the choice of dredge tailings in your pursuit of weekend prospecting as a hobby. Select an area that has produced large nuggets and one that presents easy access. Do not expect success every day. Enjoy the fresh air and scenery with your family or friends. Be especially attentive on rainy days or when the piles are wet because specimens are often washed and made visible then.

Chapter 10

BENCH TESTING

Bench testing offers a simple method of determining metallic content and monetary value of an ore specimen. Easy to conduct, it is inexpensive and safe for the beginner or weekend prospector. Knowledge of a few simple test methods and recovery systems will greatly increase your enjoyment of this hobby.

Many volumes have been written concerning various test methods. They can involve complicated mathematics as well as laboratory sciences such as chemistry and physics as well as mining and geology. We obviously cannot even begin to scratch the surface in explaining test procedures, even if readers possessed the necessary scientific expertise.

Furthermore, using many different methods to test the same specimen can produce inconcluisve results that baffle even experts. Only three basic methods of testing are discussed here and in the following chapter:

—Visual

—Metal detector discrimination

—Acid tests.

The first two methods are discussed in this chapter and are applicable to most specimens except black sand concentrates.

Let me emphasize that you should always seek professional help when experiencing reasonable doubt about analysis of a specimen. Learning these three basic procedures, however, will minimize the need for outside assistance and save time you might waste in testing worthless specimens.

BENCH TESTING INSTRUCTION

Your detector must be capable of correctly identifying metal conductivity in relation to nonconductive magnetic iron.

1. Use a small searchcoil.

2. Adjust the detector in the discrimination mode in accordance with manufacturer's instructions.

3. Determine the degree of conductivity in your sample.

Place the sample on a wooden bench with the coil parallel to it. Begin the test at zero discrimination and gradually increase discrimination. Responses help you decide whether your sample is basically a conductive (metallic) substance or magnetic iron.

As the discrimination control is increased, you can determine the approximate reading where your specimen is rejected. This will be a rough indication of its conductivity. Experience with various specimens will increase your faith in the reliability of this process.

Roy Lagal demonstrates bench testing with a metal detector to determine whether a sample contains conductive ore. Use of a calibrated instrument with small, sensitive searchcoi is necessary for proper ore sampling (also called high grading).

4. Pulverize the sample. Use a mortar and pestle or other method to render part of your sample into the finest form of dust possible.

5. Wash the sample with clean water and pan it in a small finishing pan, using standard panning procedures. If you find that oil or grease is contaminating the sample, use a dissolving agent to remove the grease.

6. Inspect your concentrates with a magnifying glass for visual identification. When values in the sample cannot be identified visually use more sophisticated methods described in the next chapter.

7. Clean all tools. It is important that you thoroughly wash all tools used in pulverizing and panning your sample to avoid contaminating specimens in later tests.

Bench testing of black sand concentrates is impossible to conduct with a metal detector. Magnetic content of the black sand will mask all gold values. Black sand salvaged from field operations must be reworked with controlled separating devices capable of recovering gold values.

Further bench test procedures are described in my book, "Detector Owner's Field Manual."

Chapter 11

BLACK SAND CONCENTRATES

One of the most frustrating experiences for the gold prospector is to see ultra fine gold in black sand concentrates after numerous failures to recover it by panning. As we have noted, bench testing of black sand with a metal detector is impossible because its magnetic content will mask gold values. Black sand must be processed with controlled separating devices, and such separating methods also apply to samples which could not be assayed by simpler methods described in the previous chapter.

Methods of separation range from sophisticated procedures employed on a large scale by smelters to the "homespun" techniques devised by miners. These include:

—Crushing or milling the concentrates and reclaiming by amalgamation;

—Leaching with cyanide or acid solutions;

—Electrolysis;

—Centrifugal separators: bowls, wheels, barrels and

—Vibrating tables.

Basic explanations of devices and systems applicable to the weekend prospector will now be given. Many may be too costly or appear too complex because of the volume of concentrate to be processed. Information presented in this book and that provided by manufacturers of separating equipment should enable you to select equipment and processes that will be appropriate to your operation.

Let me emphasize a warning about acid-testing kits. Several inexpensive kits are available, and they are comparatively safe and easy to use. We urge you not to attempt to perform acid tests, however, without experienced supervision until you are experienced in their use.

BLACK SAND CONCENTRATES

Black sand is simply the accumulation of several minerals, each weighing more (having a higher specific gravity) than the gravel or dirt in which they become deposited. Because mineralized areas differ widely, black sands come in many forms. Sometimes they contain great values; sometimes they are worthless. Granular sizes will also vary, depending primarily on iron deposits in the area. The black color comes from the predominance of small particles of magnetic iron ore. Grey sands, such as hematite and limonite contain iron but, unless heated, this sand is non-magnetic and defies efforts to remove it with a magnet.

A variety of other minerals can be found in black sand concentrates, including gold, silver, platinum, tungsten, mercury, lead, galena, manganese and zinc,

Tommy Long digs out black sand concentrate into his Gravity Trap gold pan to examine targets. After removing detectable nuggets, he will pan remaining contents to recover fine gold. This technique of detector scanning/panning is called sniping.

along with such gemstones as garnets and sapphires and other rare earth minerals. With the exception of platinum, gold is heavier than all of these.

Being soft and malleable, however, gold is easily ground by water and wind action into fine pieces invisible to the naked eye; hence, the terms "fine" or "flour" gold. Even though gold is heavier than magnetic black sand, its sometimes flat shape makes it difficult to remove, especially by ordinary panning.

Particles of black sand are often microscopic-sized gold nuggets with magnetic iron oxidation. Not only does this crust obscure the nugget from visual indentification but it must be removed before the gold can be recovered. Removal is accomplished by crushing, followed by acid or leaching treatments.

Before gold recovery from any black sand is attempted, however, you must carefully determine if sufficient values are present in the sand to make recovery economical. The gravity separation process can be used because it is quicker and generally less expensive than other recovery processes described here.

Water must be treated with a base-cutting biodegradable detergent before it can be used with any gravity separation device. Fine gold particles, coated with oil from natural vegetation and waste, tend to float in water because of their flat shape. Do not use soap with a grease base. A biodegradable detergent such as Basic H or LOC is required. Clorox or Purex can be used in small amounts, but they are not biodegradable and thus constitute pollutants, Add just a small amount of detergent to the water for your separation process to cut grease and prevent gold flakes from floating.

Finally, you should recognize that black sand concentrates can sometimes contain your only profit from an operation. Never discard black sand until you are convinced that no further gold can be profitably extracted. Remember, some firms purchase concentrates based on their assayed value, and other operators will rerun your discarded minerals for a percentage of recovered worth.

The following methods are presented for those weekend prospectors who choose to process black sand:

CONCENTRATING BOWL

A concentrating bowl is a gravity trap device designed to separate ultra fine gold from concentrates. It is considered one of the most efficient gravity devices for separating ultra fines from screened black sand. Operation is governed by the law of inertia; i.e. a mass in motion will remain in motion, and a stationary mass will remain stationary unless acted upon by an external force. Fine gold dust stays at the bottom as water carries black sand out of the bowl. Gold fines are thus not disturbed in this gravity separation process.

Being bowl-shaped and about the size of a large kitchen basin, it is not capable of quickly processing large quantities. I should point out, however, that no device processes black sand concentrates quickly. Ultra fine gold is just too difficult to capature for recovery ever to be rapid. A concentrating bowl is probably the most efficient of all gravity type separating devices.

The bowl is constructed of either plastic or metal, with plastic being preferred. Its inside is slick, and a pipe in the middle permits reuse of water.

Any type hose or small water pump is attached to an opening in the bottom, and water is forced into the bowl. This creates a swirling motion that carries lighter black sand up and out of the bowl, leaving fine gold on the bottom. The controlled and stable action permits separation of ultra fine gold.

Just a cupful of black sand can be placed in the bowl at one time, and the concentrates must have been screened down to thirty mesh or smaller. Users of gravity type separting devices screen concentrates again with a smaller mesh after each run to produce more fines.

CONCENTRATING BOWL OPERATION

1. Set the bowl on a level water tank or other surface.

2. Dump in about one cup of concentrates screened at least as small as thirty mesh.

3. Begin water action, and adjust water to the point of flowing over the edge of the bowl. Black sand will flow up, then down the center pipe and out of the bowl, leaving fines on the bottom.

4. Use a gold suction bottle to pick up fines and discharge them into your gold bottle.

5. Screen discarded concentrates down to sixty mesh or smaller and rerun.

Very fine flour gold will remain in the black sand. To recover this dust you may need nitric acid to remove associated iron. In using acid be careful not to use it in full strength since it may dissolve the gold.

Concentrating bowls, both large and small, operate more slowly than other gravity devices. They are relatively inexpensive, however, and their operation requires no expert knowledge. Average cost for a bowl is generally less than $100, and tap water can be furnished from a garden hose or small pump.

GOLD SPINNERS/WHEELS

A gold spinner is circular in design like a wheel with grooves cut in a spiral from the outside of the circle to the center. As the wheel revolves, water is forced in a spray patter onto the grooves, and lighter black sand is discarded over the edge. Heavier fine gold is slowly carried in the continuous groove until it is deposited through the center of the wheel into a container. This process, similar to an endless belt operation, is very efficient. A system can be constructed with one wheel, or in multiples of two or more, each discharging into the next until the final wheel produces gold. The advantage of multiple designs is their ability to use unscreened material.

If you wish to separate extremely fine or flour gold, you must screen down your concentrates. Because the wheel was designed to separate fine gold from black sands, screening of concentrates and paying careful attention to the slope of the wheel and the flow of water are very important. Concentrating wheels can process sand with larger grain sizes than can be processed in a bowl, but wheels are not as efficient, especially in recovering very fine dust. Continuous discharge of water on the face of the bowl creates turbulence. The bowl process may be slower and require that concentrates be screened, but fines remain on the bottom of the bowl and are not disturbed by water action.

CONCENTRATING WHEEL INSTRUCTIONS

1. Follow manufacturer's instructions in setting the wheel at a suggested angle of discharge.
2. Fill the reservoir with "treated" water.
3. Adjust water flow onto the face of the wheel.
4. Begin operations.

The angle and slope of the wheel must be carefully adjusted to permit water to flow correctly on the face before fine gold will travel in the grooves, through the center of the wheel into your gold container.

Design of your wheel will also vary according to its manufacturer. Some are open-face in design; some have the spiral groove cut into the plastic face. Those with grooves require that sands be screened small enough so that larger pieces will not become lodged in the grooves. The face of the wheel should be kept as smooth as possible and water kept clean (degreased) at all times, especially when reprocessing material. Sands should be screened after each run and hand fed. If they have been dried, some type of gravity feed system can be used. Time spent on such small chores as cleaning and screening will be well rewarded.

Initial purchase price of a gold spinner can be four to ten times that of a small bowl. This is not unreasonable, especially in relation to the amount of concentrates recovered by a suction dredge. Even when careful panning and other methods recover all possible values, the loss of remaining ultra fines could amount to a substantial amount of revenue—especially in certain mining districts. Cost of operating a wheel in the field is small—just a few cents for detergent plus fuelling your pump with batteries or gasoline.

If your dredging or weekend prospecting produces enough concentrates to make the task of panning them overly time consuming or boring, there is no more practical choice than the concentrating wheel. Extra values you gain will probably more than offset its cost.

MECHANICAL GOLD PANS

Mechanical gold pans are motorized devices made of metal and plastic. Engineered to expedite the separation of fine gold from black sand or light gravel, they will generally handle greater volumes and larger individual pieces of material than wheels or bowls. They require only a small amount of water to operate, and their purchase price is reasonable. Of course, maintenance is required, the same as with anything mechanical.

The mechanical pan and motor-driven sluice should grow in popularity as hobbyists become aware of the values being left behind even by profitable and efficient operators. In fact, the weekend prospector's profits can sometimes be greater at a previously worked site than in new areas of unbroken gravel or rock. Old tailing piles and worked-over sands offer less resistance because of their loose condition, and they do not require the tremendous amounts of water necessary to break up or dissolve hard-packed gravel.

For the hobbyist to achieve profit from working old tailing piles or sands his reprocessing methods must have a better recovery rate than those of the original operator. Luckily, the "long toms," sluices and "rockers" used in early-day mining had deplorable recovery rates, especially with finer gold. Today, the mechanical pan or sluice device reigns supreme! Its recovery rate is generally excellent; these mechanical devices lose only the finest flour gold yet process sufficient yardage of small gravel or sand to earn a profit. Using a "grizzly" or other screening device to keep larger pieces of material from the separating process, one person can easily hand-feed the hopper. A water hose can serve as an intake pump, and it is often possible to recirculate water. Mechanical pans have been available for some time, but only with higher gold prices and wider use of plastic have they become popular. Plastic allows smooth, shaped parts to be sturdily constructed without the expense of metal-casting. The new mechanical pans are proving surprisingly effective in many locations.

Before committing yourself to any particular area, good research should be conducted concerning the success of its previous operator. You should also pan the area yourself at the lowest levels where fine gravel and sand were discarded. If you find gold values still present, and if sufficient water is available, one of the fine mechanical pans or sluices now being manufactured may let you profit from the leavings of someone else. From Nome to the Deep South there are millions of dollars in fine gold values just waiting for the operator with a mechanical device capable of recovering this abandoned treasure.

BALL MILLS/TUMBLERS

Ball mills, which are simply mechanical devices used to crush ore and black sand concentrates, come in many sizes. Large mills are used by commercial smelters. Medium to small mills are generally used by miners who crush only a few tons of ore or concentrates. Small, hand-operated mills are available for the weekend prospector to crush a few pounds of concentrates. The hand-operated mills are slow and require hand labor and close attention, but the cost of a larger mill is usually prohibitive. Any venture you plan to make into this method, beyond purchase of a small, hand-operated model, should be undertaken only with sound advice from professionals concerning potential profitability of your proposed operation.

TUMBLERS

A tumbler is a small barrel-type device used by rockhounds to achieve a finished or polished surface on gemstones. The rotating or "tumbling" action abrades the surface of material loaded in the barrel. Different abrasives can polish gemstones to almost any desired finish. The addition of liquid agents such as caustic soda, acids or detergents can remove iron oxide from the small fine gold found in black sand concentrates. Mercury can then be used to gather or "amalgamate" these fines.

Never begin the amalgamation process until the free gold and other easily recovered values have been removed by pans, wheels, bowls, sluices or other gravity devices. Of course, large commercial milling operations run vast quantities of raw ore or concentrates, but they are not to be confused with your small mining or hobby operation.

SMALL TUMBLER OPERATIONS

1. Dump the black sand or other concentrate in the tumbler on an operating surface.
2. Add caustic soda or other agent you select.
3. Operate the tumbler approximately two to four hours.
4. Inspect and clean the contents.
5. Add a small amount of mercury and run for approximately the same time or until the mercury has gathered the fine gold.

Your mercury will have broken into several globules. These can be removed by hand-panning the contents or running them through any gravity device. You can also add a small amount of additional mercury, and the small drops will gather into one large drop.

Your gold is now out of the black sand and amalgamated with mercury, where its removal is relatively simple. A mercury press will quickly remove mercury from gold. You can also squeeze the amalgamate through a dense material such as chamois skin or a gold syringe with cotton to remove most mercury. Deposit the compressed mercury still containing fine gold into a Pyrex container with a 90% nitric acid solution in clean water. (Do not use 100% acid because it will simply dissolve the ultra fine gold.) Boil the amalgamate for 10 minutes in the 90% acid solution in a clean, well-ventilated area, and the mercury will evaporate. Next, pour the acid and gold solution into a container and add water and baking soda to neutralize the acid. Pour off the solution and remove your gold. Do not pour this solution down your sink or in any sanitary sewer where it could pollute.

Other methods of separation include burning the mercury and leaving a small gold "button." Old-time prospec-

Author tests black sand to determine if gold is present. Separating fine gold from black sand requires considerable care, including careful panning and use of magnifier, magnet or one of the effective mechanical gold separation devices.

tors amalgamated their gold and burned off the mercury in a metal gold pan. Not only is this about the only method available in the field but it can be accomplished easily.

There are many "tricks" associated with the burning process. You should never get your fire too hot since this can literally evaporate your gold. When you burn the amalgamate on a piece of metal, make certain to have a sealing agent under the button to keep the gold from melting into the porous metal. Blackboard chalk will seal a metal surface, or you can use borax or other chalk compounds; even paper can be used. A favorite trick of the old prospectors was to use cigarette paper when they burned amalgamate on their gold pans. Your remaining gold button is classified as "sponge gold" and is generally 85 to 90% pure.

Finally, a potato can be used to recover mercury. Cut the vegetable in half with a hole in its flat side large enough to contain the amalgamate button and place it over the button during the burning process. the mercury will condense inside the potato but can easily be shaken loose for reuse.

Extreme caution should be observed whenever acids or mercury are being used. Be especially careful not to inhale their fumes.

Considering the tumbler one last time, it can be used for other purposes than recovering fine gold. Perhaps the least known is to reprocess large natural gold nuggets. These valuable nuggets often have a coating of various oxides that detract from their natural beauty. Trying to remove the oxides with acid can discolor the gold and further damage market value of the nuggets. Your inexpensive tumbler is the solution. Simply place the nuggets in the tumbler with two capfuls of steel shot (available at any sporting goods or mining supply store), add detergent and water and tumble overnight or longer. The nuggets will not wear down but will regain the natural color of any placer nugget just coming from a crystal clear stream. Small imperfections will have been removed and its beauty enhanced. Gold jewelry can also be treated in this manner without harm.

VIBRATING TABLES

Vibrating tables—called "jigs" because of their motion—are used in both commercial and hobby operations. Large professional jigs are precisely engineered pieces of complicated machinery. Large jigs usually try to divert the fine sand that contains values over or through the jig to recover as many values as possible without milling. More information on such large and expensive devices is available from commercial mining supply manufacturers.

The small miner is generally interested in models designed to separate black sand concentrates. These tables, generally designed either to vibrate or remain stationary, operate with a controlled flow of water, and they are quite capable of recovering extremely fine gold. For some reason these small tables have never achieved the popularity of wheels or bowls. Perhaps it is because they resemble a common sluice box. Also, they are generally more difficult to install, and they are time-consuming to operate. If you are interested in a table-type separating device, I suggest that you observe them in action and evaluate your needs carefully before proceeding further.

SPIRAL SLUICE BOXES

A spiral sluice box is essentially one half of a circle. They are not in the traditional square or rectangular shape of a box and do not resemble the standard sluice. Imagine a length of ordinary culvert such as found under a roadbed, only made of plastic. Halve it and you have a spiral sluice. Indentations are designed in a continuous groove pattern similar to the concentrating wheel. With a precisely controlled flow of water, their construction allows these spiral sluices to process unscreened black sands. Water flow is critical, and the discharge angle is particularly important. Your gold's "travel" through the sluice must be observed at all times. Spiral sluice boxes can be efficient, but they will not actually duplicate the action of concentrating wheels or bowls.

GRAVITY DEVICES

In concluding this discussion of gravity-type separation devices I have tried to settle some of the confusion concerning their efficiency and application. Let me emphasize that all of these devices are being used successfully today by professionals in the fields of gold discovery and processing. They have their differences which can be quite important to the weekend prospector:

– Some require a large area for assembly.

– Some do not process unscreened or unclassified material.

– Some operate more rapidly than others but recover a lower percentage of ultra fines.

– Some require large amounts of water that cannot be recycled.

Regardless, any of the gravity devices, if operated properly, will show some profit because the black sand concentrates being processed would have been discarded or abandoned. Therefore, your separating device, whether purchased or designed, will contribute to your success. In fact, the separation of fine gold values from heavy black sand concentrates challenges weekend prospectors today as one of their greatest areas for potential profit. Using modern separation devices and methods can let you become highly successful!

LEACHING METHODS AND ACIDS

Leaching processes and acids should be used only after all possible gold fines have been recovered by gravity methods. Commercial operations commonly use cyanide to recover values from ore that has already been processed. Because cyanide is one of the most dangerous poisons known to man, you should never use or even experiment with it unless accompanied by a professional who is competent in its use. Actually, the cyanide process is a dangerous procedure that has no value to the small miner or prospector. You may have some curiosity in learning how large mills use cyanide to "leach" or remove values from ore, then recover them by another process. If so, there are numerous volumes devoted to this subject.

Leaching processes that do not require cyanide are available to the weekend prospector. They are designed to release the gold into a solution where it can be recovered by such methods as electrolysis. Leaching is relatively easy to perform but does not leave your fine gold in as natural a state as amalgamation or the mercury process. Therefore, we do not recommend that you seriously consider leaching as a solution to your separation needs. We urge that you consider these needs carefully in regard to the scope and size of your operation before you decide to use a leaching process. Of course, materials can then be purchased from many different manufacturers who will provide extensive information on their effectiveness.

Acids have been used to recover gold for many years, but I have always promoted safer methods of separation. I do not offer instructions to the weekend prospector on the use of acids. Why? Because acids are deceptively dangerous. Disasters can occur in an instant, particularly involving innocent bystanders or children. The novice should use acids only accompanied by a person experienced in their use. When you become proficient enough to use acids safely, you will no longer be a novice!

If proper precautions are employed, including complete neutralization after its use, nitric acid can be the safest acid for a prospector to use. Nitric acid does not affect gold, and it dissolves only the oxidized iron attached to gold without attacking attached quartz or rock. HydroFLUoric acid generally dissolves rock attached to gold ore specimens or nuggets after heating for several hours. HydroCHLoric acid is combined with nitric acid to form aqua regia (king's water) which dissolves gold into a liquid. (Three parts of hydrochloric acid are required to one part of nitric acid in the solution.) Both hydroFLUoric and hydroCHLoric acid are extremely dangerous.

Nitric, hydrofluoric and hydrochloric acids are those generally used in processing gold. We have intentionally been brief concerning information about them and have offered few instructions on their use. All acids are dangerous, some more than others. Just a single drop of hydrofluoric acid, for example, can burn a hole in flesh. When heated, acids often produce deadly gases. Acid and cyanide will combine to form a gas that is not only deadly but impossible to smell. No matter what your knowledge of chemistry, be extremely cautious in using acids to process gold.

Books relating to jewelry-making and casting generally cover the treatment of gold and acid very thoroughly. Generally the use of safe gravity separation procedures and limited use of nitric acid will fill the needs of the weekend prospector or hobbyist.

Chapter 12

SAFETY PRECAUTIONS

Follow the Eleventh Commandment: BE CAREFUL.

Weekend prospecting will take you far away from smog, city traffic, street muggings and the chance of other city incidents. Hazards of a different sort will appear from time to time in the gold fields, and I urge you to heed the various warnings given in preceding chapters. The most dangerous hazards you may face—even moreso than those involving the use of acids—are old mines and tunnels. Since most accidents involving cave-ins are fatal, I urge you to follow all of my safety precautions and to add a few of your own.

Some old mines and shafts contain deadly gases. Since many of these gases are difficult to detect by smell alone, their presence will be unknown until it is too late for escape. Extreme care should be taken, therefore, in deep excavations of any kind. Remember the old miners with their canaries? If the canary died, poison gas was present; if not, the mine was safe. Hard on the small birds, perhaps, but lives were saved. Often gas extinguished a miner's carbide lamp, but this was never a foolproof detection method. At other times gases would explode at the first spark of an open flame. Battery-powered lights were an absolute necessity. Let me repeat: if poisonous or incendiary gas is present in an old mine and remains undetected, the weekend prospector can expect fatal results. Never enter an open shaft or mine without experienced professionals present. Remember the Eleventh Commandment.

Snakes have not been discussed, but any reasonable person tries to avoid them. If bitten, don't panic and run. Try to remain as calm as possible and get help. The old "cut-and-suck-the-poison" theory has been thoroughly debunked. You can do more harm cutting at a bite than the snake inflicted originally. Ice applied to the bitten area tends to slow down circulation and is recommended by many doctors. Tourniquets applied between the bite and the heart will help, but they must be loosened frequently to prevent blood poisoning.

I've been rattlesnake bitten twice, and I can assure you that it's hard NOT to panic. The sickness that comes immediately after the bite may convince you that death is a viable alternative. Just get medical help as soon as possible and hang in there. Few people die from snake bites in the United States, and those who do generally die from sheer panic or from problems they brought on themselves.

Wild animals such as bears, bobcats and skunks inhabit old mines and caves. When entering an unexplored area, look out for animals and don't get in their way. Remember that their main aim is to get out of your way. The only harm that wild animals generally cause is the fright they give you. The skunk, of course, is an animal of a different stripe. All your running and dodging and yelling is generally to no avail. Just pay your money and take your chances with this creature. A lot more prospectors have been skunk-sprayed than snake-bitten.

Scorpions, spiders and small insects such as wasps and bees seldom produce more than immediate pain and discomfort. Since their bites can sometimes be lethal or debilitating, however, care should be taken with them at all times. Some individuals have allergies that can transform a minor insect bite into a major medical problem. At any rate, when you go out into the field, watch out for bugs of all kinds.

My travels into foreign countries have revealed the presence of countless untold hazards, both natural and manmade. Knowledge of and a healthy respect for snakes, insects and the old mine tunnels kept me safely away from most of those dangers, and I have always tried to avoid bandits and other human varmints as much as possible.

Weekend prospecting can be a glorious hobby for men and women of all ages. You can enjoy yourself in the great outdoors with fresh air, exercise and the chance to see new and beautiful scenery. Just be careful and don't spoil it with a careless accident.

Conclusion

You have now been exposed, through the written word, to the many aspects of weekend prospecting presented in this book. A lifetime of experience would be necessary for you to become an expert in just one of the recovery methods presented herein. This is true regardless of the intelligence or experience of the individual. Field experience and great amounts of study are required to learn any phase of prospecting.

By touching lightly on each subject and intentionally keeping instructions to a minimum, I have tried to present an overall picture to the weekend prospector. I hope that the beginner now understands the basic procedures of prospecting. At the same time, I hope those professionals reading this volume may have been alerted to some hitherto unknown prospecting method or piece of equipment.

My entire lifetime has been spent in the prospecting/treasure hunting field. Several of my books on electronic prospecting methods have become accepted as international professional guides. It would have been impossible for me alone, however, to have assembled all the knowledge necessary for this book without the help of several well known experts. To them, I express deep gratitude for their kind help. For you weekend prospectors, I profoundly hope that this instructional book will guide you to countless years of enjoyment and much material wealth.

Glossary

AUTOMATIC VLF GROUND ELIMINATION (CANCELING)—A type of metal detector circuit that requires no manual adjustments to achieve iron/earth/mineral elimination (canceling). Circuits continually analyze the soil beneath the searchcoil and automatically adjust the detector circuitry to "ignore" the minerals. Do not confuse Automatic VLF Ground Elimination with Automatic (Audio) Tuning since they have two entirely different characteristics.

BLACK SAND—See Magnetic Black Sand.

CHIMNEY—A column of rock/minerals created when the earth's internal heat and pressure forced molten ore up through cracks and fissures. The ore cooled, solidified and became exposed when the surrounding softer earth materials eroded.

CLASSIFICATION, AUDIO—An audio method(s) of classifying detected targets into conductivity classes or categories.

CLASSIFICATION, VISUAL—A visual (metered or light) method(s) of classifying detected targets into conductivity classes or categories.

CONCENTRATE—Heavier metal(s) that remain after lighter sands have been washed away or otherwise were eliminated.

DISCRIMINATOR—The capability of a detector to identify ranges or detected targets. See Ground Elimination.

FALSE DETECTION—Responses to objects or anomalies other than sought after targets.

FERROUS—Pertaining to iron and iron compounds.

FLOAT—Ore which has broken away from its Mother Lode and "floated" or moved away from the point of break.

GROUND ELIMINATION (Also CANCELING)—

The ability of a metal detector to eliminate (ignore, cancel) the detection effect of iron minerals.

HEADPHONES—Device to convert electrical energy waves into audible waves of identical form. Especially useful for metal detector users working in noisy or windy locations. They will do everything loud speakers will do and generally do it better. In addition, they are less susceptible to damage by rain and require much less power than a speaker.

HOT ROCK—A mineralized rock containing either a greater or lesser amount of magnetic iron than the search area that produces a positive (but false) signal in a metal detector.

MAGNETIC BLACK SAND—Magnetite, a magnetic oxide of iron, can also contain titanium and other rare earth minerals. Serve mainly as an indicator for the possible presence of placer gold.

MATRIX—The entire area area below a searchcoil which is illuminated by the transmitted electromagnetic field. A matix may wholly, partially or intermittently contain conductive and/or non-conductive (ferrous and/or non-ferrous) materials such as salts, sulfides and metallic ores.

METAL DETECTOR—An electronic instrument or device capable of sensing the presence of conductive objects lying within the earth or otherwise out of sight and of providing the operator with an audible or visual indication of that presence.

METAL—As related to operation of a metal detector, a ferrous or non-ferrous substance capable of conduction electricity and producing a positive response from the detector.

METAL/MINERAL—As used to describe a metal detector, means that a detector has detection characteristics and/or the ability to distinguish between the two.

MINERAL—As related to operation of a metal detector, only magnetic iron in its native state, i.e. black sand or iron oxides, affect the electromagnetic field of a metal detector in a negative (undesirable) manner.

MODE—The manner of operating a metal detector, generally at the operator's option; i.e. search modes, metal or mineral modes, VLF all-metal modes, trash elimination modes.

NON-FERROUS—Pertains to metals and minerals without any iron; i.e. brass, silver, lead, aluminum.

PENETRATION—Ability of a detector to penetrate earth material or air, wood, rock and water to locate metal targets. Penetration is a function of detector design for the material being penetrated.

PLACER—Pronounced plasser (plaster without the "T"). An accumulation of gold, black magnetic sand and other elements of specific gravity heavier than accompanying sand and rock. Generally occurring in low spots or streams on the downside of boulders and other obstructions.

POCKET—An accumulation of ore in a concentrated area.

RIFFLE—The indentations found in a gold pan. Should be 90 degrees to be effective.

SCANNING—Movement of the detector searchcoil over the area being searched. Also called "sweeping."

SUPER-SNIPING—Trademarked detecting method that uses a small-diameter (3-4") searchcoil to enhance individual target detection by eliminating nearby target influence. Especially valuable in trash areas.

TARGET—The sought-for object which is distinct and apart from matrix anomalies. A coin is a target but foil from a gum wrapper is not. Junk iron is not normally a target, yet for bottle hounds using detectors to locate old trash pits it is. A target is any sought object lying within a matrix that may or may not contain other but unsought conductive objects and which, by virtue of its material makeup, is capable of being located with an electronic metal detector. The term is also used to describe all detectable metal or mineral targets.

TR—Abbreviation for "transmitter-receiver," which is a type of metal detector circuit.

TUNING—Manual adjustment made by an operator to bring the detector's sound to the desired audible threshhold. Some detectors incorporate a circuit that automatically (and electronically) returns the audio to threshold.

VLF—Very Low Frequency, which is a segment of the radio frequency spectrum that includes frequencies from 3 kilohertz to 30 kilohertz. It also denotes a type of detector.

VLF MODE—The term identifying the VLF operational setting on a metal detector.

Please send the following books

- ☐ **Treasure Recovery from Sand and Sea $12.95**
- ☐ **Buried Treasure of the United States . . . . $ 9.95**
- ☐ **Weekend Prospecting . . . . $ 3.95**
- ☐ **Find an Ounce of Gold a Day . . . . $ 1.00**
- ☐ **Successful Coin Hunting . . . . $ 8.95**
- ☐ **Modern Metal Detectors . . . . $ 9.95**
- ☐ **Detector Owner's Field Manual . . . . $ 8.95**
- ☐ **Treasure Hunter's Manual #6 . . . . $ 9.95**
- ☐ **Treasure Hunting Pays Off! . . . . $ 4.95**
- ☐ **VLF-TR Metal Detector Handbook . . . . $ 8.95**
- ☐ **Electronic Prospecting . . . . $ 4.95**
- ☐ **Gold Panning Is Easy . . . . $ 6.95**
- ☐ **Treasure Hunter's Manual #7 . . . . $ 9.95**

Garrett Guides to Treasure

- ☐ **Find An Ounce of Gold a Day $1.00**
- ☐ **Metal Detectors Can Help You Find Wealth $1.00**
- ☐ **Find Wealth on the Beach $1.00**
- ☐ **Metal Detectors Can Help You Find Wealth $1.00**
- ☐ **Find Wealth in the Surf $1.00**

Please add 50¢ for each book ordered (to a maximum of $2) for handling charges.

Total for Items	$__________
Texas Residents Add 8% Tax	$__________
Handling Charge	$__________
Total of Above	$__________
MY CHECK OR MONEY ORDER IS ENCLOSED.	$__________

I prefer to order through

☐ VISA or ☐ MasterCard (Check one.)

____

Card Number

____

Expiration Date

____

Phone #

____

Signature (Order must be signed.)

____

NAME

____

ADDRESS

____

CITY, STATE, ZIP

Ram Publishing Company
P.O. Drawer 38649
Dallas, TX 75238